SUIFAN
Squared Paper

Made in the USA
Las Vegas, NV
08 October 2022

56811057R00057